Orlando Austin New York San Diego Toronto London

Visit *The Learning Site!*
www.harcourtschool.com

Introduction

How do you find your way home from a friend's house? You find your way by looking for things that you know. Houses, streets, trees, signs—all of these can help you find your way.

Now suppose that you are a sailor on a ship from long ago. Your ship is about to sail out on the ocean.

There are no buildings or streets. There is no land at all—just wind and ocean water. How will you find your way to where you are going?

Traveling by Sea

People have sailed the oceans for thousands of years. Some went to explore. Others did so to trade. Others wanted to find treasure in new lands.

Sailors had to navigate to get where they were going. Navigation is steering a ship on a course.

Sometimes sailors would steer close to the coastline. They drew maps to help them. They followed currents in the water. They watched for animals in the water and in the air.

But what about trips across the ocean? There was no land to guide them. So sailors learned to navigate by the sky and the stars instead.

The Night Sky

Not everyone sees the same stars in the sky at night. The stars you see depend on where you are on Earth.

There is an imaginary line that divides Earth into two parts, the northern half and the southern half. This line is called the equator.

People who live north of the equator can see many stars. They can see the brightest star in the sky, Sirius. They can also see Polaris, known as the North Star.

People who live south of the equator can see other stars. They can see a bright group of stars called the Southern Cross.

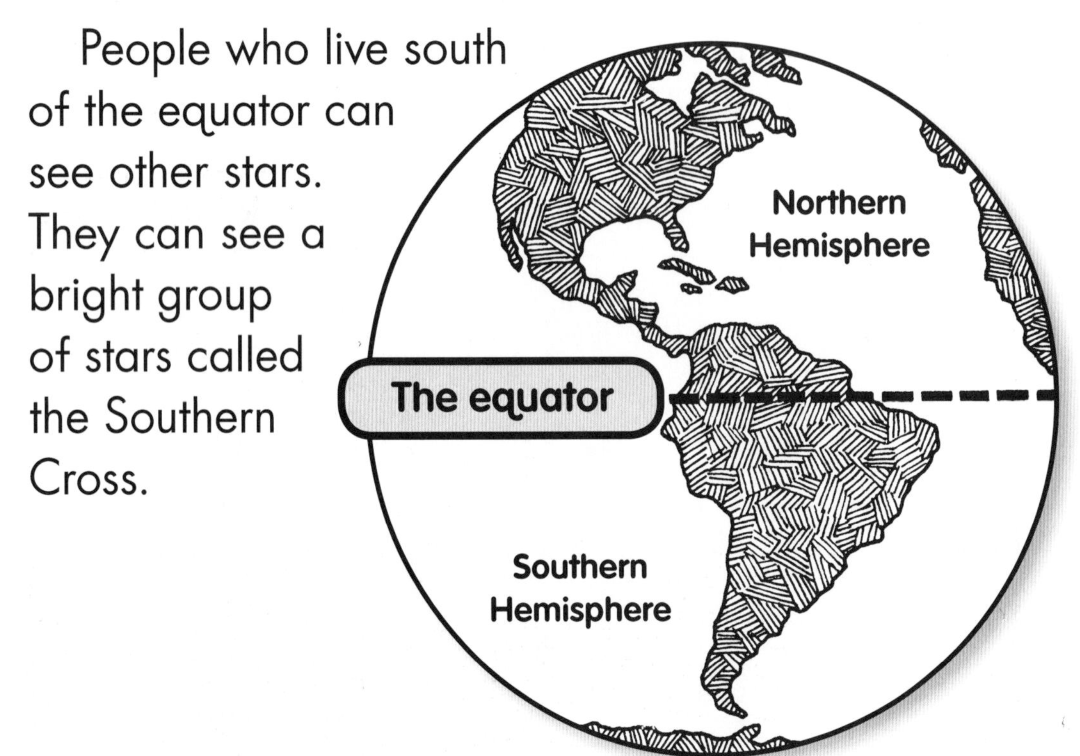

The North Star

If you stand at the North Pole, the North Star would be almost directly above you. The North Star is also called the pole-star.

North of the equator, the North Star guided sailors. When sailors saw the North Star in the sky, they knew which way was North.

The North Star also helped sailors know where their ship was located between the North Pole and the equator. This position is called latitude.

Sailors measured the angle from the North Star to the horizon (where the earth or water meets the sky). This angle was their latitude.

Finding latitude

How to Find the North Star

The North Star is not the brightest star in the sky. It is not always easy to find.

There is a way to help you spot the North Star. Look for a group of stars shaped like a large cup with a long handle. This group of stars is known as the Big Dipper.

Find the two stars at the front of the cup. These two stars are sometimes called the Pointer Stars. Follow the Pointer Stars upward. The next bright star you see will be the North Star. It is in the Little Dipper group of stars.

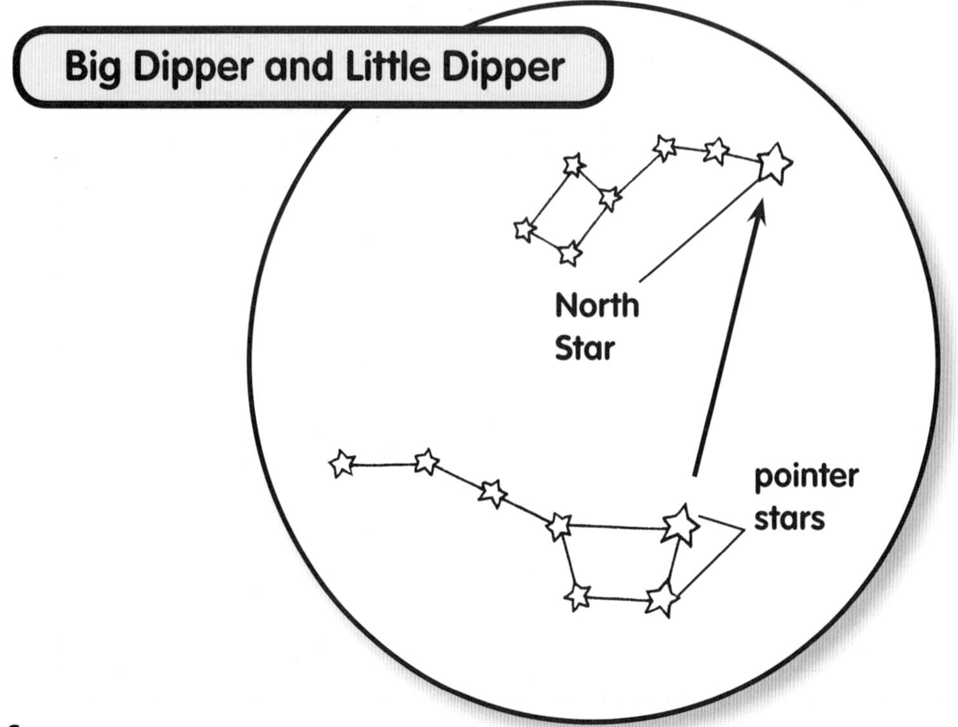

Tools for Navigation

Sailors used many tools to navigate. One of the earliest was the quadrant. The quadrant had two straight edges set at a right angle. An arc joined the edges. The arc was marked to show degrees. A weighted line hung from the top of the quadrant.

The sailor would line up the North Star with sighting holes in the quadrant. The weighted line would fall on one of the arc's degree marks. This was the ship's latitude.

The sextant was another tool used by sailors. The sextant used small mirrors. It measured the angle of a star or the sun above the horizon.

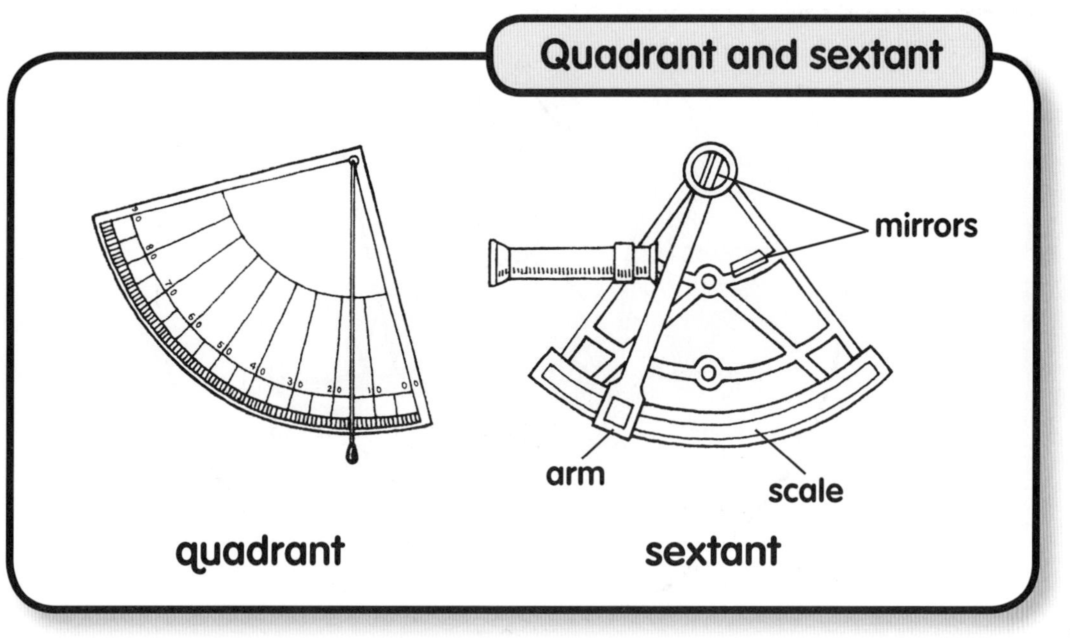

Wayfinders

The early sailors of the Pacific Ocean islands did not use tools, such as sextants. Over many years, they learned to navigate by reading signs from nature. They looked for ocean swells and currents. They studied the sun and the stars. They made star charts. By doing all of these things, they found the way for others who followed. That is why they became known as "wayfinders."

The wayfinders and sailors of long ago were brave. From them, we have learned much about the oceans and how to find our way by the stars.

Wayfinders in canoe